Integrative Anatomy and Pathophysiology in Traditional Chinese Medicine Cardiology

Integrative Anatomy and Pathophysiology in Traditional Chinese Medicine Cardiology

Dr. Anika Niambi Al-Shura, BSc., MSOM, Ph.D
Continuing Education Instructor
Niambi Wellness
Tampa, FL

Medical Illustrator: Samar Sobhy

ELSEVIER

AMSTERDAM • BOSTON • HEIDELBERG • LONDON
NEW YORK • OXFORD • PARIS • SAN DIEGO
SAN FRANCISCO • SINGAPORE • SYDNEY • TOKYO
Academic Press is an imprint of Elsevier

Academic Press is an imprint of Elsevier
32 Jamestown Road, London NW1 7BY, UK
The Boulevard, Langford Lane, Kidlington, Oxford, OX5 1GB, UK
Radarweg 29, PO Box 211, 1000 AE Amsterdam, The Netherlands
225 Wyman Street, Waltham, MA 02451, USA
525 B Street, Suite 1900, San Diego, CA 92101-4495, USA

British Library Cataloguing-in-Publication Data
A catalogue record for this book is available from the British Library

Library of Congress Cataloging-in-Publication Data
A catalog record for this book is available from the Library of Congress

ISBN: 978-0-12-800123-3

For information on all Academic Press publications
visit our website at **store.elsevier.com**

This book has been manufactured using Print On Demand technology. Each copy is produced
to order and is limited to black ink. The online version of this book will show color figures
where appropriate.

Working together
to grow libraries in
developing countries

www.elsevier.com • www.bookaid.org

DEDICATION

The energy and effort behind the research and writing of this textbook is dedicated to my son, Khaleel Shakeer Ryland. May this inspire and guide you through your journey in your medical studies, career, and life.

ACKNOWLEDGMENTS

This is a special acknowledgment to my seven-year medical students at Tianjin Medical University (2012–2013) who served as cardiovascular research assistants. May your future medical careers be successful.

An Qi He
Bin Lin Da
Han Jiang
Chen Hua
Jia Ying Luo
Jun Zhang
Lin Lin
Ming Lu
Nang Zhang
Ping Tang
Hu Si Le
Zhao Tian Man
Wen Xing Ning
Xing Wen Zhao
Tang Ying Mei
Li Ying Ying
Xiong Yong Qin
Ding Yu
Li Yan Jun

CONTENTS

SECTION III PATHOLOGY

INTRODUCTION

The companion course for this textbook edition can be found on the Elsevier website and at www.niambiwellness.com.

APPROVING AGENCIES

The medical continuing education course which uses this textbook is called, Integrative Anatomy and Patho-physiology in TCM Cardiology. The course is approved by the National Certification Commission for Acupuncture and Oriental Medicine (NCCAOM) as course #1053-002 for 14 PDA points, and by the Florida Board of Acupuncture as course #20-334890 for 15 CEUs.

COURSE DESCRIPTION

This course examines of the structure, function, and pathology of the heart from the Western medicine and Chinese medicine perspectives.

COURSE OBJECTIVES

- Review the normal anatomy and physiology of the cardiovascular system.
- Evaluate the disease characteristics found in parts of the heart sections.
- Analyze the interrelationship between Western medicine and TCM theory in cardiology.

Cardiovascular Anatomy

CHAPTER 1

Cardiac Cells

CHAPTER OBJECTIVES

After studying this chapter, you should be able to:

- List the five main features of cell physiology.
- Describe the significance of myocytes and calcium Ca++ release.
- List the four features of vascular functioning.
- Describe the vascular cells and their function.
- Describe vasodilation and constriction.
- Describe the function of endothelial cells.

1.1 PART 1: CELL PHYSIOLOGY

1.1.1 Lesson 1: Cell Physiology
1.1.1.1 Myocytes
These cells are found in cardiac, skeletal, and smooth muscles, and they contain the protein troponin. Cardiac myocytes generate the electrical impulses throughout the heart. Bundles of myofilaments, called actin and myosin, make up the sarcomeres and are responsible for contractility (muscle contraction).

1.1.1.2 Excitation–Contraction Coupling
Excitation–contraction coupling, the muscle tissue stimulus and response mechanism, depends on a process that induces calcium release. Calcium ions (Ca++) are released by the sarcoplasmic reticulum to initiate the energy needed for various cardiac systems to function.

1.1.1.3 Inotropy
Inotropy is the strengthening/weakening function of the cardiac tissue.

- Positive = calcium release causes strengthening.
- Negative = hypercalcemia or weakening.
- H + (acidosis) is also a negative inotrope.

Integrative Anatomy and Pathophysiology in TCM Cardiology. DOI: http://dx.doi.org/10.1016/B978-0-12-800123-3.00001-3

1.1.1.4 Lusitropy
- Lusitropy is the reuptake of calcium for cardiac relaxation.
- Positive = chemical manipulations assisting the process through the cyclic adenosine monophosphate (cAMP) pathway.
- Negative = drugs which block calcium release (inotropy).

1.1.1.5 Metabolism
- Cardiac myocyte contractility is constant, requiring a lot of energy through ATP.
- ATP is produced aerobically.
- Large amounts of mitochondria are found within myocytes.

1.2 PART 2: VASCULAR FUNCTION

1.2.1 Smooth Muscle Cell
Smooth muscle is the tissue of all vascular walls. These include the aorta, arteries, arterioles, veins, and lymphatic vessels.

1.2.2 Vasodilation
- Catecholamine binding to beta-adrenergic receptors
- Carbon dioxide exposure

1.2.3 Vasoconstriction
- Norepinephrine and epinephrine mediated through alpha 1-adrenergic receptors
- Oxygen exposure

1.2.4 Endothelial Cell
These cells line the interior of all vessels throughout the cardiovascular system and function to protect the vessels:

- Contains heparin sulfate for antithrombosis
- Controls inflammation
- Controls blood pressure

NOTES

For chapter 1 tutorial log on at www.niambiwellness.com to access the companion course for module 1.

CHAPTER 2

Cardiac Structure

CHAPTER OBJECTIVES

After studying this chapter, you should be able to:

- Describe the sections of the heart and vascular system.
- List the cardiac arteries and function.
- Describe the layers and function of a vein.
- Describe the function of capillaries.
- Describe the function of the lymphatic system.

2.1 PART 1: HEART ANATOMY

Go to chapter 2 in the online course to find the answers in the chart

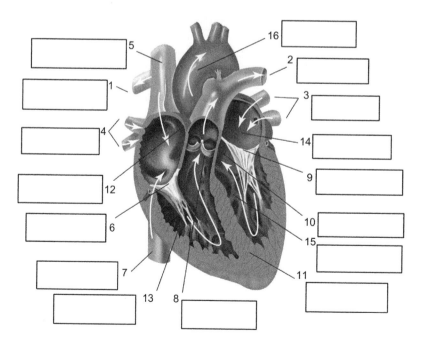

Integrative Anatomy and Pathophysiology in TCM Cardiology. DOI: http://dx.doi.org/10.1016/B978-0-12-800123-3.00002-5
© 2014 Elsevier Inc. All rights reserved.

Section of the heart	Function
1. Right pulmonary Artery	This vessel carries blood, with low levels of oxygen, from the right ventricle to the lungs.
2. Left Pulmonary Artery	This vessel carries blood, with low levels of oxygen, from the right ventricle to the lungs.
3. Left Pulmonary Veins	This vessel carries oxygenated blood from the lungs to the left atrium of the heart.
4. Right Pulmonary Veins	This vessel carries oxygenated blood from the lungs to the left atrium of the heart.
5. Superior Vena Cava	This vessel receives blood from the head, arms and chest and sends it into the right atrium.
6. Atrial Ventricular (AV)/ Tricuspid Valve	This valve allows blood to pass from the right atrium to the right ventricle.
7. Inferior Vena Cava	This vessel is formed from the two iliac veins. It receives blood from lower limbs and abdominal organs and sends it into the right atrium.
8. Pulmonary Valve	This valve is also called the semilunar valve. It opens to send blood from the right ventricle through the pulmonary artery towards the lungs.
9. Biscupid/ Mitral Valve	Allows blood to pass between the left atrium and the left ventricle.
10. Aortic Valve	This valve opens to allow blood flow from the left ventricle through the aorta.
11. Septum	This is the wall that separates the left and right ventricles.
12. Right Atrium	This chamber receives blood from the vena cava.
13. Right Ventricle	This chamber receives blood from the right atrium and pumps it into the pulmonary artery.
14. Left Atrium	This chamber receives blood from the pulmonary veins.
15. Left Ventricle	This chamber that receives blood from the left atrium to pump into the aorta.
16. Aorta	This artery facilitates movement of blood from the left ventricle back to the heart, out to all limbs and organs, excluding the lungs.

2.2 PART 2: CARDIAC ARTERIES

Go to chapter 2 in the online course to find the answers in the chart

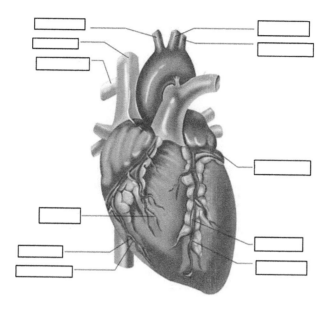

Artery	Function
Right Coronary Artery (RCA)	This artery supplies the right side of the heart.
Posterior Interventricular Artery (PIV) Patent Ductus Arteriosus (PDA)	The PIV can be a branch of the circumflex coronary artery, also a branch of the left coronary artery. The PDA is a branch of the right coronary artery RCA.
Marginal Artery	This artery is a major branch of the right coronary artery (RCA).
Left Coronary Artery (LCA)	This artery comes up from the cusp of the aortic valve, then bifurcates as the left main artery into the anterior interventricular artery and the left circumflex artery (LCX).
Left Circumflex Artery(LCX)	This artery curves to the left towards the posterior surface of the heart and joins the right coronary artery. It supplies the left ventricle and the papillary muscle.
Anterior Interventricular Artery (LAD)	This artery runs behind the pulmonary artery, comes forward to reach the anterior interventricular sulcus then continues down to the apex.

2.3 PART 3: VEINS

Veins are smooth muscle tissue and also contain elastic and fibrous tissues. They typically have a diameter of about 5 mm and wall thickness of about 0.5 mm.

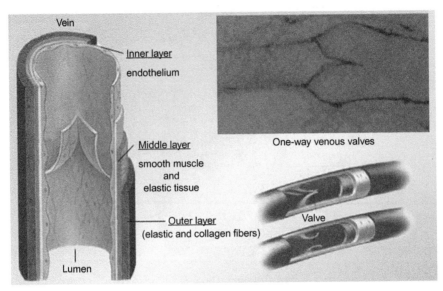

Image: from creative commons

Venous section	Function
Tunica intima	This section is the inner layer which includes a sheet of elastic fibers and covers the endothelial cells.
Tunica media	This section is the middle sheet layer which includes a mixture of smooth muscle and elastic fibers.
Tunica externa	This section is the outer layer which includes a sheet of elastic and collagen fibers.

Use chapter 2 in the online course to lable the chart

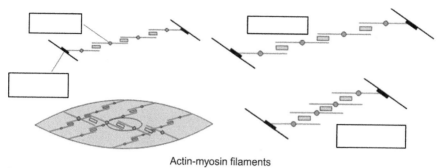

Actin-myosin filaments

Image: from creative commons

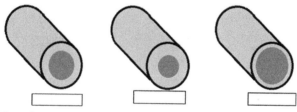

Image: from creative commons

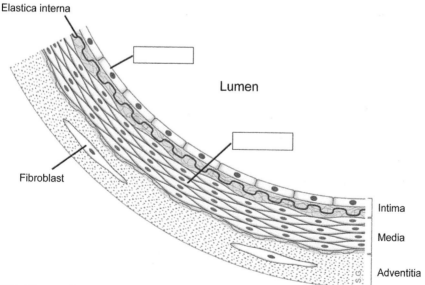

Elastica interna

Lumen

Fibroblast

Intima

Media

Adventitia

Image: from creative commons

2.4 PART 4: CAPILLARIES

Use chapter 2 in the online course to lable the chart

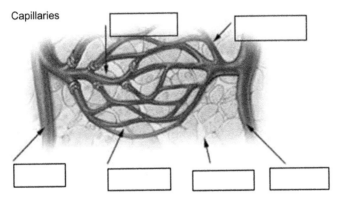

Image: from creative commons

The walls of the capillary do not include smooth muscle tissue but do have a single layer of endothelial cells that are about 0.5 mm thick.

There are also contractile elements that allow flexibility and response to mediators.

Capillaries have low velocity, allowing for the exchange of nutrients and waste between regular circulation and the interstitial fluid surrounding the cells.

2.5 PART 5: LYMPHATIC SYSTEM

Use chapter 2 in the online course to lable the chart

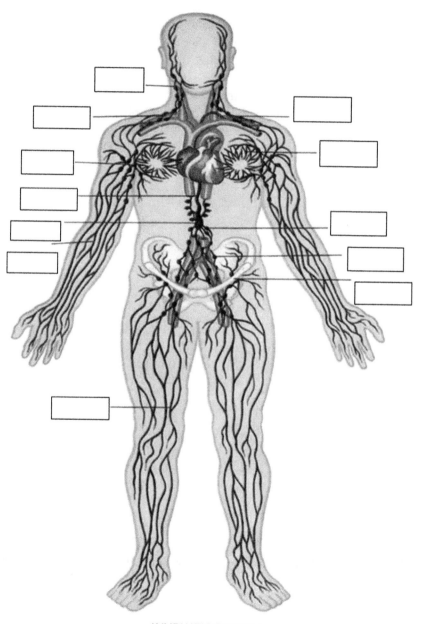

LYMPHATIC SYSTEM

In the case of a normal "textbook" body, 8 L of water is returned to blood circulation through the lymphatic system.

2.5.1 Functions of the Lymphatics

- Absorption of digested fats in the gut
- Return of leaking plasma proteins from around interstitial spaces back into the circulation
- Tissue drainage and fluid distribution
- Defense center housing the lymphocytes and phagocytes

NOTES

For chapter 2 tutorial log on at www.niambiwellness.com to access the companion course for module 1.

CHAPTER 3

Cardiac System

CHAPTER OBJECTIVES

After studying this chapter, you should be able to:

- Trace a drop of blood through the circulatory system.
- Name the main artery connecting the heart zang to the trunk organs.
- List all of the arteries and veins connecting the heart zang to the zang organs.
- List all of the arteries and veins connecting the heart zang to the fu organs.

Integrative Anatomy and Pathophysiology in TCM Cardiology. DOI: http://dx.doi.org/10.1016/B978-0-12-800123-3.00003-7

3.1 PART 1: HEART CIRCULATION

Use chapter 3 in the online course to lable the chart

CIRCULATION OF BLOOD THROUGH THE HEART

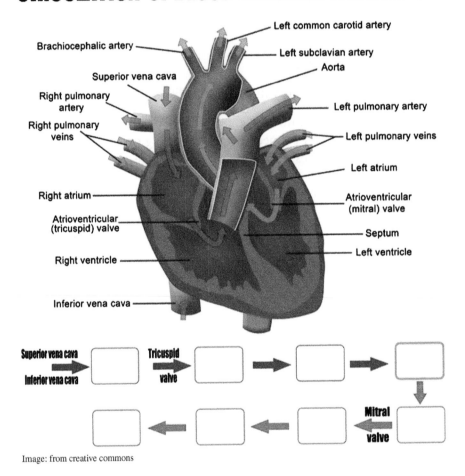

Image: from creative commons

3.2 PART 2: SYSTEMIC CIRCULATION

3.2.1 Heart Zang to Brain Sui
3.2.1.1 The Blood Supply to the Anterior Portion of the Brain
- Internal carotid arteries
- Anterior cerebral artery (ACA)

• Middle cerebral artery (MCA)

3.2.1.2 The Blood Supply to the Posterior Cerebral Portion of the Brain: Occipital Lobes, Cerebellum, and Brainstem

• _____
• _____
• _____
• _____
• _____
• _____
• _____
• _____

3.2.2 Heart Zang to Arms

• Subclavian artery
• Axillary artery
• Brachial artery

3.2.3 Heart Zang to Gastrointestinal Areas

3.2.3.1 Heart Zang to Stomach Fu and Spleen Zang

From the abdominal aorta:

• _____
• _____
• _____

Venous drainage:

• _____
• _____
• _____
• _____
• _____

3.2.3.2 Stomach Fu to Small Intestine Fu

Duodenum service:

• _____
• _____

3.2.3.3 Supply to Intestine Fu
The superior and inferior mesenteric artery supplies the sections of the ileum, cecum, large intestine fu, and appendix:

• Ileocolic artery

3.2.3.4 Heart Zang to Intestine Fu
From the abdominal aorta:

• Superior mesenteric artery

   Venous drainage:

• Right colic artery (ascending colon)
• Middle colic artery (transverse colon)
• Left colic artery (descending colon)

3.2.3.5 Heart Zang to Liver Zang
From the abdominal aorta:

• _____
• _____

   Entrance into the liver zang:

• Central vein of each lobule
• Exiting the liver
• Hepatic veins, which leave the liver and return the flow back to the heart.

3.2.4 Heart Zang to Kidney Zang
From the abdominal aorta:

• _____

Branching = _____ > _____ > _____

NOTES

For chapter 3 tutorial log on at www.niambiwellness.com to access the companion course for module 1.

Module Review Questions

1. List the five main features of cell physiology.
2. Describe the significance of myocytes and calcium Ca++ release.
3. List the four features of vascular functioning.
4. Describe the vascular cells and their function.
5. Describe vasodilation and constriction.
6. Describe the function of endothelial cells.
7. Describe the sections of the heart and vascular system.
8. List the cardiac arteries and function.
9. Describe the layers and function of a vein.
10. Describe the function of capillaries.
11. Describe the function of the lymphatic system.
12. Trace a drop of blood through the circulatory system.
13. Name the main artery connecting the heart zang to the trunk organs.

14. List all of the arteries and veins connecting the heart zang to the zang organs.
15. List all of the arteries and veins connecting the heart zang to the fu organs.

Log on at www.niambiwellness.com to access the companion course and quiz for Module 1.

Cardiovascular Physiology

CHAPTER 4

Cardiac Rhythm

CHAPTER OBJECTIVES

After studying this chapter, you should be able to:

- Describe actin and myosin in cardiac myocytes.
- Discuss the role of ATP for synthesizing qi energy within myocytes.
- Explain electrical activity resting and action potentials, and signaling transmission.

4.1 PART 1: CARDIAC MUSCLE STRUCTURE

The contractile unit, or bundle, on a myocyte is called the *sarcomere*. It is defined by the Z-line, which is known as the protein a-actin. Under polarized light microscopy, A and I bands that are attached to the sarcomere can be seen.

- A = myosin
- I = actin

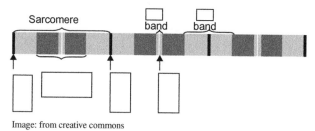

Image: from creative commons

Please visit the online course to record the answers on the figure.

The sarcomeres store Ca++ and are in bundles that lie side by side to make up the myofibril. Myocytes are cardiac muscle cells that contain many myofibrils, a cell nucleus, and mitochondria, which is a site of qi activity.

Integrative Anatomy and Pathophysiology in TCM Cardiology. DOI: http://dx.doi.org/10.1016/B978-0-12-800123-3.00004-9

4.2 PART 2: CARDIAC MUSCLE CONTRACTILITY

Increase in intracellular Ca++ causes the sliding action of the actin and myosin, which contracts the myocyte. Ca++ binds to troponin C and becomes part of a three-troponin protein complex. The cross bridge between the actin and myosin undergoes cycle changes with muscle contraction. Each cycle involves hydrolysis, or using up, of an ATP (qi energy) molecule as part of the contraction. The fibers of the muscles derive energy from ATP (adenosine triphosphate) through oxidative phosporylation and contain a rich amount of mitochondria.

4.2.1 Ca++ Regulation

T tubules are very close to the sarcoplasmic reticulum (Ca++ storage), and they release Ca++ to help spread electrical activity throughout the myocyte. During muscle relaxation, ATPase sends some Ca++ back to the sarcoplasmic reticulum. Stimulation of the sympathetic heart nerves (b-adrenoceptor) causes an increased contraction and an increase in the second messenger cAMP. ATP makes cAMP and creates pathways to assist certain hormones such as adrenaline and glucose, which normally have trouble entering cells, to enter and activate energy within a cell. Phosphodiesterase is an enzyme in cAMP, which degrades signaling effects over time. Caffeine and other stimulants can prolong the effect of cAMP.

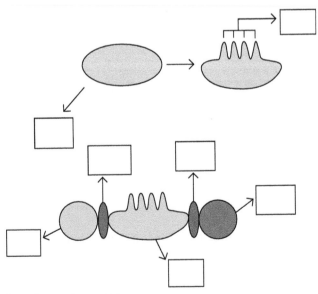

Image: from creative commons

Please visit the online course to record the answers on the figure.

4.3 PART 3: ELECTRICAL ACTIVITY

4.3.1 Resting Potential

The balance of ionic concentrations of Na+ and K+ is important. The diffusion of K+ assists in the resting phase of action potentials in myocytes. Positive ionic balance is explained through the Nernst equation. The Goldman equation considers negative ions such as Cl− movement across cell membranes.

4.3.2 Action Potentials

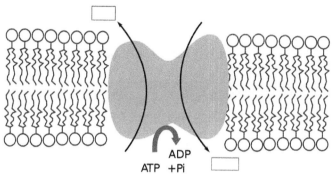

Image: from creative commons

Please visit the online course to record the answers on the figure.

$$E_{m,\,K_x\,Na_{1-x}\,Cl} = \frac{RT}{F}\ln\left(\frac{P_{Na^+}[\quad]_{out} + P_{K^+}[\quad]_{out} + P_{Cl^-}[\quad]_{in}}{P_{Na^+}[\quad]_{in} + P_{K^+}[\quad]_{in} + P_{Cl^-}[\quad]_{out}}\right)$$

Please visit the online course to record the answers on the figure.

$$E = \frac{RT}{zF}\ln\frac{[\text{ion}\qquad\text{cell}]}{[\text{ion}\qquad\text{cell}]} = 2.303\,\frac{RT}{zF}\log_{10}\frac{[\text{ion}\qquad\text{cell}]}{[\text{ion}\qquad\text{cell}]}$$

Please visit the online course to record the answers on the figure.

The resting potential of myocytes is −85 mV when the cell is excited and triggers the opening of the sodium gates.

This action potential in ventricular tissue has five phases:

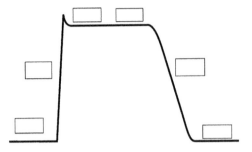

Please visit the online course to record the answers on the figure.

Phase	Action
Phase 0	Sodium gates open when PNA permeability is greater than Potassium (K+).
Phase 1	K+ permeability/ PK increases, leaving the cell at a favorable electrical gradient. Also the opening of Ca++ channels causes inflow of calcium from outside the cell.
Phase 2	The calcium plateau is reached.
Phase 3	The membrane potential decreases in value.
Phase 4	The cycle concludes back to the resting potential.

4.3.3 Pacemaker Tissue

The ability of the cardiac muscle tissue for spontaneous depolarization and action potential is called automaticity. The sections that have automaticity are:

- SA node
- AV node
- Bundle of His and the Purkinje fibers

Heart rates from the SA node	Cardiovascular status
110—120 bpm	This is the healthy "textbook" SA node during active period.
70 bpm	The resting heartbeat is under the parasympathetic innervation.
50 bpm	Failure of the SA functioning places more stress on the AV node. Failure of the AV node is known as complete heart block
30—40 bpm	The ventricles have the ability to beat on their own, with the help of purkinje fibers.

4.3.4 Transmission

Transmission happens in the SA node because it is the site of the fastest rate. The AV node controls the electrical impulses between the atria and ventricles, to allow blood to flow properly between them. The depolarization passes to the bundle of His and the Purkinje fibers to contract the right and left ventricles in order to send blood out of the body. This is all reflected in the ECG deflections.

NOTES

For chapter 4 tutorial log on at www.niambiwellness.com to access the companion course for module 2.

CHAPTER 5

Pumping Actions

CHAPTER OBJECTIVES

After studying this chapter, you should be able to:

- Discuss heart qi and blood movement as represented through the six phases of the Wiggers diagram.
- Discuss the significance of end diastolic volume, end systolic volume, and stroke volume.
- Describe the factors involved in filling pressure that determine the amount of blood entering the ventricles during diastole.

5.1 PART 1: CARDIAC CYCLE

5.1.1 The Heart Governs Blood

According to Traditional Chinese Medicine (TCM), the main function of the heart is governing production and circulation of blood. The heart houses spirit, which means that the heart has a relationship with the brain. The relationship is mainly about physiological functioning and manifests within mental and psychoemotional activities. It is also noticed as spirit for balanced functioning for the rest of the zang and fu organs. This spirit is also called heart qi.

5.1.2 Heart Spirit is Heart Qi

Physiologically, heart qi becomes the energy from the autonomic nervous system and transforms it to control overall cardiovascular functioning. This includes the circulation of the blood outside of the heart through the vessels, and the control of dilation and constriction of the vessels. The heart qi also controls the smoothness, pressure, and the rate of blood flow. The activity of the heart governing of blood is explained in stroke volume. The characteristics of heart qi are explained through the cardiac cycle and regulation. The measurements are calculated through electrocardiogram (EKG) and blood pressure, felt on the pulse, and heard in auscultation.

Integrative Anatomy and Pathophysiology in TCM Cardiology. DOI: http://dx.doi.org/10.1016/B978-0-12-800123-3.00005-0

5.1.3 Wiggers Diagram/Heart Qi

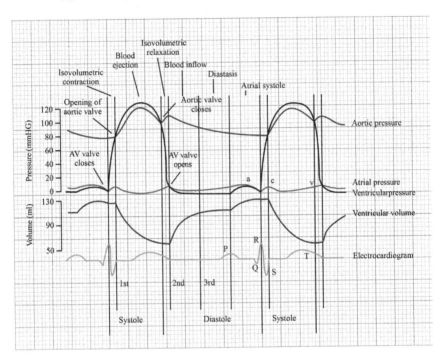

The cardiac cycle includes the events related to the flow and pressure of blood through the heart from one heartbeat to the next. The cycle is represented through six phases:

5.1.3.1 Phase 1: Atrial Systole

The P wave represents the atrial electrical qi depolarization. This phase is ventricular *diastole*. During filling, pressure within the right atrium increases, pushing blood across the AV valves into the right ventricle. At the end of the phase, the ventricles are completely filled to about 140 mL. This is the end *diastolic* volume (EDV).

If a S4 (fourth heart sound) is the sound heard during this phase, it is usually a sign of ventricular hypertrophy.

1. Where is phase 1 located on EKG?

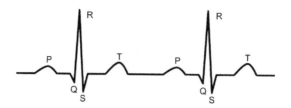

Go to the course and circle the EKG chart according to the lesson.

5.1.3.2 Phase 2: Isovolumetric Ventricular Systole

The QRS wave, on the electrocardiogram section of the Wiggers Diagram (image 5.1) represents the beginning of systole. This phase represents ventricular qi depolarization. The rise in pressure in the ventricles exceeds the pressure in the atrium, causing the AV valves to close. The S1 (first heart sound) is the sound heard during this phase.

2. Where is phase 2 located on EKG?

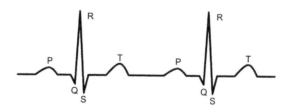

Go to the course and circle the EKG chart according to the lesson.

5.1.3.3 Phase 3: Ventricular Ejection

This section involves the S and T waves. When pressure in the ventricles exceeds the pressure within the pulmonary arteries and the aorta, the pulmonic valves open to allow blood flow. No heart sound is heard in healthy valves. A sound heard at this phase is called an ejection murmur.

3. Where is phase 3 located on EKG?

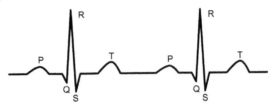

Go to the course and circle the EKG chart according to the lesson.

5.1.3.4 Phase 4: Ventricular Relaxation

This section represents the T wave after the QRS wave. The pressure in the ventricle decreases due to emptying. Atrial pressure is rising. No heart sound is heard in healthy valves. A sound heard at this phase is called an ejection murmur.

4. Where is phase 4 located on EKG?

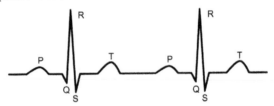

Go to the course and circle the EKG chart according to the lesson.

5.1.3.5 Phase 5: Isovolumetric Relaxation

This section represents the end of the T wave. The volume of blood that remains in the left ventricle, which is equal to 70 mL, is called the end systolic volume (ESV).

5. Where is phase 5 located on EKG?

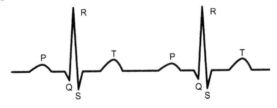

Go to the course and circle the EKG chart according to the lesson.

EDV − ESV = SV (140 − 70 = 70). S2 sound is heard as the valve closes.

5.1.3.6 Phase 6: Diastolic

This section represents the end of the T wave to the middle of the P phase. The atria fills with blood, and the pressure causes the AV valves to open to release the blood into the ventricles. No sound is heard in healthy AV valves. If an S3 sound is heard, it is considered normal in children but is ventricular dilation in adults.

6. Where is phase 6 located on EKG?

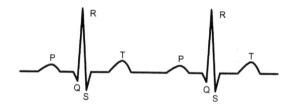

Go to the course and circle the EKG chart according to the lesson.

5.2 PART 2: REGULATION

This is the heart function of governing blood volume to guarantee enough to nourish all tissues and organs of the body.

The Rule of 70:
EDV − ESV = SV

The textbook normal heart pumps 5 L of blood throughout at 70 bpm. When the ventricle empties, the volume drops from 140 mL EDV to 70 mL ESV, with 70 mL as stroke volume (SV).

This is the TCM heart relationship with the brain, where the marrow of the brain transmits qi energy to the heart. The heart spirit controls the transmitted energy as heart qi, which regulates physiological functioning.

5.2.1 Regulation of Heart Rate

$$\text{Heart rate} \times \text{stroke volume} = \text{cardiac output}$$

The rate of the beating heart is determined by the sinoatrial node (SAN). The central nervous system (CNS): sympathetic and parasympathetic nervous system is responsible for functioning. The CNS controls heart rate and baroreceptor reflex.

Sympathetic fibers on the right side of the body have the effect on heart rate; sympathetic fibers on the left side of the body are responsible for contractility. Noradrenaline/norepinephrine is released at the nerve endings. It continues to act on the beta 1 receptors of the SA node and increases Ca^{++}. Through the parasympathetic innervation, the right vagus nerves mainly go to the SA node, and the left vagus nerves mainly go to the AV nodes. The parasympathetic release, the neurotransmitter acetylcholine, acts upon muscarinic (M2) receptors, and also through the G protein helps to reduce cAMP production in SA node cells. Cyclic AMP (cAMP) activates $Na+$ and $Ca++$ channels, causing depolarization of the pacemaker cells. This decreases heart rate.

5.2.2 The Fick Principle

$$Q = \frac{V_{O_2}}{C_A - C_V}$$

Log into the companion course for the lesson on the Fick principle.

5.3 PART 3: PRELOAD ON STROKE VOLUME

The TCM heart functioning of governing blood volume is seen in the role of blood volume and venous tonicity. Filling pressure determines the amount of blood entering the ventricles during diastole.

5.3.1 Frank–Starling Mechanism

This law states that the work done by the myocardium in system is related to the resting length of muscle fibers in diastole.

5.3.2 Blood Volume

Atrial natriuretic peptide (ANP) is released into the contained blood when the walls of the atrium are stretched. ANP increases glomerular filtration rate in the kidneys and is important in the renin—angiotensin system. The renin—angiotensin system regulates the sodium and water system, which is part of the regulation of blood volume. The reduction of blood pressure or blood volume promotes the secretion of renin, which is influenced by:

• Sympathetic nervous system
• Baroreceptor reflex

5.3.3 Venomotor Tone

The amount of blood traveling in the vessels depends upon the amount of blood volume and the ability of the vessel walls to be stretched. The sympathetic nervous system acts on the veins through the $\alpha 1$ adrenoceptors. The effects are seen as gravity on hydrostatic pressure, upon moving from a lying or sitting to a standing position. A drop in cardiac output and arterial blood pressure can cause dizziness and fainting. In addition, a patient with a fast heart rate who is also losing blood, thus reducing blood volume, can suffer hemorrhagic shock.

5.4 PART 4: INOTROPY (CONTRACTILITY)

5.4.1 Increase

The left branch of the sympathetic nerve innervation to the $\beta 1$ adrenoceptor of the ventricles increases contractility. Contractility has to do with the stretching of muscle fibers, during which changes happen to the amount of $Ca++$ in the myocyte.

5.4.2 Decrease

$H+$ ion is the strongest negative inotrope. CO_2 into cardiac myocytes leads to weakening contractions. Sympathetic nerve innervation through β receptors on the myocytes is connected to cAMP formation. The result is an influx of $Ca++$ and increased reuptake in the sarcoplasmic reticulum. Systole is shortened and diastole can be prolonged, which benefits blood flow through the coronary arteries. Channel blockers can manipulate cAMP mediated pathways, in other words, beta blockers and calcium channel blockers.

5.5 PART 5: AFTERLOAD ON STROKE VOLUME

5.5.1 Left Ventricle

When the left ventricle wall exceeds pressure, the aortic valve opens, and blood is ejected into the aorta. Pressure in the aorta will therefore affect the workload on the ventricle. Additional factors include the chamber radius and wall thickness. Increased pressure leads to reduction in stroke volume, which in turn leads to cardiac failure.

NOTES

For chapter 5 tutorial log on at www.niambiwellness.com to access the companion course for module 2.

CHAPTER 6

Vascular Function and Circulation

CHAPTER OBJECTIVES

After studying this chapter, you should be able to:

- Explain the significance of $Ca++$ and $H+$ on smooth muscle contraction.
- Describe the vasodilation factors found in endothelial tissue.
- Discuss the significance of nitric oxide (NO) on vascular tissue.

6.1 PART 1: CALCIUM CHANNELS

The relaxation of smooth muscle tissue requires reduction of intracellular $Ca++$. The acid–base status also has an influence on the amounts of calcium. $H+$ ions (acidosis) can displace calcium on binding sites, causing an increased need for $Ca++$ ions. While diagnosing diseases, this is the rationale for clinical assessments of total calcium.

6.2 PART 2: CONTRACTION OF SMOOTH MUSCLE

Contraction of smooth muscle includes $Ca++$, actin, and myosin release but does not include troponin. The $Ca++$ released during contraction binds to calmodulin, which promotes phosphorylation of myosin. When $Ca++$ amount decreases, the myosin is also dephosphorylated yet retains its attachment with actin, with slow gradual detachment with as little ATP consumption as possible.

6.3 PART 3: ENDOTHELIAL FACTORS

6.3.1 Endothelins

Endothelins ET-1, ET-2, and ET-3 are a group of peptides. ET-1 is involved with vasoconstrictors such as angiotensin and is important in regulatory effects in cardiovascular, kidney, and neurological functioning.

Integrative Anatomy and Pathophysiology in TCM Cardiology. DOI: http://dx.doi.org/10.1016/B978-0-12-800123-3.00006-2

It is important to note that endothelins are also involved in gene expression and protein synthesis in these organs.

6.3.2 Nitric Oxide

Nitric oxide (NO) is synthesized by cells that are exposed to:

- Interferon alpha (INFα)
- Tumor necrosis factor alpha (TNFα)
- Interleukin 1β (ILB1β)

NO is triggered by increase in shear stress on blood vessel walls due to an increase in arterial blood pressure and also by agonists in the receptor cell such as bradykinin, acetylcholine, and thrombin. The angiotensin converting enzyme (ACE) activates bradykinin and ACE II. NO increases the production of cGMP (a secondary messenger like cAMP, but it is involved in smooth muscle relaxation).

6.3.3 Prostaglandin

Prostaglandin PGI2 released from the endothelial cells inhibits platelet aggregation. It also acts as a vasodilator and increases the production of cAMP.

6.4 PART 4: METABOLITES

Accumulation of metabolites in tissues is related to an increase in vasodilation. CO_2 and $H+$ are dangerous to brain activity. Adenosine and $K+$ are important in cardiac tissue.

NOTES

For chapter 6 tutorial log on at www.niambiwellness.com to access the companion course for module 2.

CHAPTER 7

Humeral Control

CHAPTER OBJECTIVES

After studying this chapter, you should be able to:

- List the hormones involved in pathogenic issues and explain the significance of serotonin and bradykinin on vascular tissue.
- Describe the role of renin–angiotensin and its involvement in heart failure.
- Describe the two adrenal hormones and their role in vasoconstriction and contractility.

7.1 PART 1: LOCAL HORMONES

These hormones are part of pathogenic issues:

- Bradykinin
- Histamine
- Serotonin
- NO
- PGI2

7.1.1 Bradykinin

This peptide-like serotonin and histamine is part of the pain response. It dilates arterioles and is part of the inflammatory response. It also binds to endothelial cells and increases the production of NO. ACE increases the half-life of bradykinin and inactivates it.

7.1.2 Serotonin

Serotonin causes vasoconstriction of large arteries and veins and contributes to some inflammatory responses. In the brain, it may be responsible for migraine headaches.

Integrative Anatomy and Pathophysiology in TCM Cardiology. DOI: http://dx.doi.org/10.1016/B978-0-12-800123-3.00007-4

7.2 PART 2: SYSTEM HORMONES

7.2.1 Renin—Angiotensin

Renin is secreted by the kidneys and generates *angiotensin I*. Angiotensin I is the precursor to angiotensin II (Ang II), which is produced in the presence of ACE. Ang II actions include:

- Increase in antidiuretic hormone (ADH) secretion
- Formation of ET-1
- Dispogenic thirst responses in the brain
- Sympathetic nervous system activities

The most important to note is the role of Ang II in fibrotic tissue change as seen in remodeling during heart failure.

7.2.2 Adrenal Hormones

The adrenal glands produce catecholamine, which consists of:

- Adrenaline (epinephrine)
- Noradrenaline (norepinephrine)

Catecholamines are also released by the sympathetic nervous system.

Catecholamine receptors are divided into two types: α and β. The one found on blood vessels is $\alpha 1$ for vasoconstriction. The one on the heart is β for contractility.

NOTES

For chapter 7 tutorial log on at www.niambiwellness.com to access the companion course for module 2.

CHAPTER 8

Nervous System Regulation

CHAPTER OBJECTIVES

After studying this chapter, you should be able to:

- Describe the location and function of the autonomic nervous system.
- Describe where the sympathetic nerves pass and explain their involvement with the cardiovascular system.
- Describe where the parasympathetic nerves pass and explain their involvement with the cardiovascular system.

8.1 PART 1: AUTONOMIC NERVOUS SYSTEM

The autonomic nervous system is found in the CNS. The section of the brain that controls respiratory functions and cardiovascular activity is the medulla oblongata of the lower brain stem. There are two branches of the autonomic nervous system:

- Sympathetic nervous system
- Parasympathetic nervous system

8.2 PART 2: SYMPATHETIC REFLEXES

These nerves pass along the thoracic and lumbar T1-L2. The sympathetic nervous system is mediated by a1 releasing adrenaline and noradrenaline for vasoconstriction.

Q: What two actions do the sympathetic reflexes do to the cardiopulmonary system?

```
┌─────────────────────────────────────┐
│                                     │
└─────────────────────────────────────┘
┌─────────────────────────────────────┐
│                                     │
└─────────────────────────────────────┘
```

Integrative Anatomy and Pathophysiology in TCM Cardiology. DOI: http://dx.doi.org/10.1016/B978-0-12-800123-3.00008-6

8.3 PART 3: PARASYMPATHETIC REFLEXES

These nerves originate from the cranial sections III, VII, IX, and X and from the sacral S2-4. The parasympathetic system is responsible for vasodilation.

What two actions do the parasympathetic reflexes go to the cardio-pulmonary system?

NOTES

For chapter 8 tutorial log on at www.niambiwellness.com to access the companion course for module 2.

Module Review Questions

1. Describe actin and myosin in cardiac myocytes.
2. Discuss the role of ATP for synthesizing qi energy within myocytes.
3. Explain electrical activity resting and action potentials, and signaling transmission. Discuss heart qi and blood movement as represented through the six phases of the Wiggers diagram.
4. Discuss the significance of end diastolic volume, end systolic volume, and stroke volume.
5. Describe the factors involved in filling pressure that determine the amount of blood entering the ventricles during diastole.
6. Explain the significance of $Ca++$ and $H+$ on smooth muscle contraction.
7. Describe the vasodilation factors found in endothelial tissue.
8. Discuss the significance of NO on vascular tissue.
9. List the hormones involved in pathogenic issues and explain the significance of serotonin and bradykinin on vascular tissue.
10. Describe the role of renin−angiotensin and its involvement in heart failure.
11. Describe the two adrenal hormones and their role in vasoconstriction and contractility.
12. Describe the location and function of the autonomic nervous system.
13. Where do the sympathetic nerves pass, and what is their involvement with the cardiovascular system?
14. Where do the parasympathetic nerves pass, and what is their involvement with the cardiovascular system?

Log on at www.niambiwellness.com to access the companion course and quiz for Module 3.

This also concludes the Integrative Anatomy and Pathophysiology in TCM Cardiology course. It is strongly suggested that you log onto the courses at the companion websites to review the course modules. Next, submit course documents and complete the final exam.

Upon passing the exam, you will receive completion certificates that include your name and practice license number, along with the specific number of credit hours awarded for this course. Electronic transmission of CEU and PDA credits will be sent to NCCAOM and your state medical board.

Pathology

CHAPTER 9

Overview of Western and TCM Perspectives

CHAPTER OBJECTIVES

After studying this chapter, you should be able to:

- Explain the direct relationship between the heart zang and other zang organs to assist in the generation, storage, and circulation of essence, qi, and/or blood.
- Explain the direct relationship between the heart zang and fu organs to assist in the generation, storage, and circulation of essence, qi, and/or blood.
- Explain the direct relationship between the heart zang and fu organs to assist in the generation, storage, and circulation of essence, qi, and/or blood.
- Explain the strength of the intrinsic factors in the blood through each organ circulation.
- Explain the effect of the extrinsic factors in the blood through each organ circulation.

9.1 PART 1: TRADITIONAL CHINESE MEDICINE

The heart zang relationship with other vital zang and some fu organs is about balancing yin and yang. This is dependent upon:

- The generation, storage, and circulation of essence, qi and/or blood.
- The balance and interaction of mental and psychoemotional activity.

Integrative Anatomy and Pathophysiology in TCM Cardiology. DOI: http://dx.doi.org/10.1016/B978-0-12-800123-3.00009-8

Heart and vicera	Generation and circulation	Psychoemotional balancing
Heart and lungs	Very strong for qi	Moderate
Heart and brain	Moderate for qi	Strong
Heart and kidneys	Strong for essence and spirit change into qi	Weak
Heart and liver	Very strong for blood circulation	Strong
Heart and spleen	Strong for generating and circulating blood	Weak
Heart and gastro-intestines	Strong for generating and transforming essence and circulating blood	Weak

9.2 PART 2: BASIC MEDICAL SCIENCE

- Vital body organs operate under certain mechanisms that are both extrinsic and intrinsic.
- The cardiovascular involvement with other vital organs is primarily mediated through the vessels. Certain extrinsic and intrinsic factors are included in pathology.
- Extrinsic factors = hormones and sympathetic nervous system involvement
- Intrinsic factors = autoregulation of adjusting and maintaining blood pressure levels, endothelial substances, and other metabolites

Circulation	Intrinsic factors	Extrinsic factors
Coronary circulation	Very weak sympathetic control	-Metabolic control -Auto-regulation
Pulmonary circulation	Very weak sympathetic control	-Very weak metabolic control -No auto-regulation
Cerebral circulation	Very weak sympathetic control	-Metabolic control -Auto-regulation
Renal circulation	Strong sympathetic stimulation	-Very weak metabolic control -Strong auto-regulation
Hepatic circulation	Strong sympathetic control	-Very weak metabolic control -No auto-regulation
Splenic circulation	Strong sympathetic control when included with the gastrointestinal system	-Moderate metabolic control -No auto-regulation
Gastrointestinal circulation	Strong sympathetic control	-Metabolic control -No auto-regulation

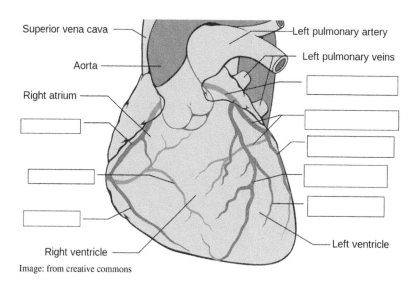

Superior vena cava —

Aorta —

Right atrium —

Right ventricle —

Left pulmonary artery

Left pulmonary veins

Left ventricle

Image: from creative commons

Please go to the online course to find the answers.

NOTES

For chapter 9 tutorial log on at www.niambiwellness.com to access the companion course for module 3.

CHAPTER 10

Lung Diseases

CHAPTER OBJECTIVES

After studying this chapter, you should be able to:

- Describe the generating and circulating properties of the lung zang.
- Describe the balancing and interacting properties of the lung zang.
- Explain the lung relationship with the heart according to Western medicine.
- Explain the TCM and Western medicine perspectives in pathology.

10.1 PART 1: TRADITIONAL CHINESE MEDICINE

The lungs produce pectoral qi by blending fresh air with the essence from food digestion. This qi becomes important for the entire body. The qi then mixes with received blood from the cardiovascular system. The lungs send the qi back into the heart, which transforms into heart qi. Blood circulation through the vessels is required to assist both lung and heart qi, and to propel the qi to the rest of the body.

The lungs dredge and regulate the circulation and excretion of water.

The circulation and excretion of water assists the head and face and also transports downward to assist the kidneys.

Generate and Circulate	Balance and Interact
Very strong for generating and circulating qi	-Moderate for psychoemotional balancing

The functions of the lungs:

- Governing qi
- Connecting with vessels
- Smoothing the water passages

Integrative Anatomy and Pathophysiology in TCM Cardiology. DOI: http://dx.doi.org/10.1016/B978-0-12-800123-3.00010-4

10.2 PART 2: BASIC MEDICAL SCIENCE

10.2.1 Pulmonary Circulation

The main purpose of the pulmonary circulation is for blood gas exchange of CO_2 and O_2 at the alveoli. The right ventricular output determines pulmonary blood flow and vascular resistance. The sympathetic nerves innervate the pulmonary veins but have a very weak effect on vascular resistance and fluid pressure. Endothelin reactive oxygen mechanisms and calcium may be involved in arterial hypoxia, which leads to vasoconstriction.

10.2.1.1 The Relationship with the Heart

- The pulmonary system includes the pulmonary and bronchial blood supply.
- The pulmonary artery supplies blood to alveoli.
- The aorta supplies blood to the trachea and other bronchial structures.

Please go to the online course to find the answers.

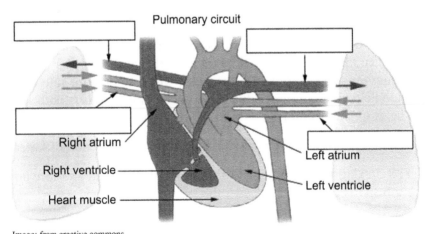

Pulmonary circuit

Right atrium
Right ventricle
Heart muscle
Left atrium
Left ventricle

Image: from creative commons

10.3 PART 3: LUNG DISEASES

Extrinsic Factors	Intrinsic Factors
Very weak sympathetic control	-Very weak metabolic control -No autoregulation

10.3.1 Balance and Interact

Emotional factors, such as grief, sorrow, and worry, consume the qi and moisture of the lungs.

10.3.2 Pathology

In pathology, the lungs are affected by physiological factors. Physiologically, if lung qi is deficient, then it fails to assist the heart for the circulation of blood. Symptoms can include chest distress, breathlessness or hyperventilating, coughing, sputum, lassitude, and low voice.

Diseases include:

- Asthma
- Bronchitis
- Chronic obstructive pulmonary disease (COPD)
- Cor pulmonale
- Emphysema
- Cystic fibrosis
- Pneumonia

When the water passageways are weak or not smooth, then there is stagnant dampness in the lung or chest and rattling phlegm.

Diseases include:

- Interstitial lung disease (ILD)
- Pulmonary edema
- Pleural effusion

When the circulation through the vessels is unsmooth or deficient, there may be palpitations and cyanosis.

Diseases include:

- Pulmonary embolism
- Pulmonary hypertension

NOTES

For chapter 10 tutorial log on at www.niambiwellness.com to access the companion course for module 3.

CHAPTER 11

Brain Diseases

CHAPTER OBJECTIVES

After studying this chapter, you should be able to:

- Describe the generating and circulating properties of the brain sui.
- Describe the balancing and interacting properties of the brain sui.
- Explain the relationship between the brain and heart according to Western medicine.
- Explain the TCM and Western medicine perspectives in pathology.

11.1 PART 1: TRADITIONAL CHINESE MEDICINE

The brain sui is considered the sea of marrow. It dominates life activities by harmonizing all organ systems. It dominates mental acuity by housing intelligence and reason. It dominates the senses through communication with the limbs and organ systems in order to maintain movement.

Generate and circulate	Balance and interact
Moderate for generating and circulating qi	Strong for psychoemotional balancing

The functions of the brain:

- Dominating life
- Dominating mental acuity
- Dominating the senses

11.2 PART 2: BASIC MEDICAL SCIENCE

11.2.1 Cerebral Circulation

A steady supply of oxygen is necessary for consumption. The brain also constantly regulates CO_2 levels as part of pressure and blood flow.

Integrative Anatomy and Pathophysiology in TCM Cardiology. DOI: http://dx.doi.org/10.1016/B978-0-12-800123-3.00011-6

This autoregulation occurs as the brain shifts pressure constantly in order to protect from the effects of elevated pressure. Metabolic control is significant to direct brain functioning. So, sympathetic activation mostly serves to assist autoregulation.

11.2.2 The Relationship with the Heart

The cerebral circulation receives about 14% of cardiac output and consumes about 20% of overall total oxygen in the body.

The brain has four main arteries:

- Left and right vertebral arteries, which join on the pons to form the basilar artery
- Left and right carotid arteries, which top the basilar artery and interconnect with smaller branches to form the Circle of Willis.

Please go to the online course to find the answers.

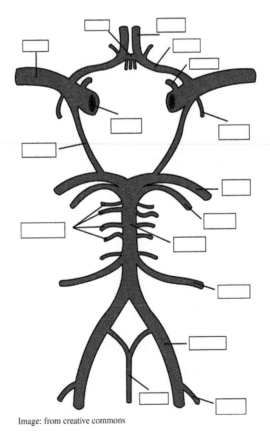

Image: from creative commons

11.3 PART 3: BRAIN DISEASES

Extrinsic factors	Intrinsic factors
Very weak sympathetic control	-Moderate metabolic control -Moderate autoregulation

11.3.1 Balance and Interact

Emotional factors are considered as part of consciousness: self-identity, personal beliefs and awareness, responsiveness, empathy and sympathy.

11.3.2 Pathology

In pathology, the brain is affected by physiological factors and depends on strong intrinsic factors for survival.

Diseases include:

- Cerebral thrombosis
- Intracerebral hemorrhage
- Stroke
- Transient ischemic attack (TIA)
- Cystic fibrosis
- Pneumonia

If blood flow to the brain is slow to block, the result is listlessness and unconsciousness.

Diseases include:

- Cerebral edema
- Ischemic stroke

NOTES

For chapter 11 tutorial log on at www.niambiwellness.com to access the companion course for module 3.

CHAPTER 12

Kidney Diseases

CHAPTER OBJECTIVES

After studying this chapter, you should be able to:

- Describe the generating and circulating properties of the kidney zang.
- Describe the balancing and interacting properties of the kidney zang.
- Explain the relationship between the kidneys and heart according to Western medicine.
- Explain the TCM and Western medicine perspectives in pathology.

12.1 PART 1: TRADITIONAL CHINESE MEDICINE

The kidney yin and yang maintain interdependence for normal metabolism and physiological activity. The kidney essence includes both congenital and acquired. Congenital comes from the parents, and the acquired from food essence from spleen and stomach. The kidney essence is the basis for kidney qi, which motivates function, energy, and development of the body through the ages. The kidney qi also functions in regulating water metabolism and separating the clear from the turbid. San jiao assists in movement of turbid fluid from the body to the kidneys. The kidneys work with the spleen in transforming and transporting clear and turbid. The clear is sent through the spleen to the lungs for regulation and circulation. The turbid is sent to the bladder as urine.

Generate and circulate	Balance and interact
Strong for essence and cultivating spirit in qi	Weak for psychoemotional balancing

Copyright © 2014 Anika Niambi Al-Shura. Published by Elsevier Inc. All rights reserved.

Integrative Anatomy and Pathophysiology in TCM Cardiology. DOI: http://dx.doi.org/10.1016/B978-0-12-800123-3.00012-8

The functions of the kidneys:

- Storing essence
- Governing water
- Receiving and governing qi.

12.2 PART 2: BASIC MEDICAL SCIENCE

12.2.1 Renal Circulation

The kidneys need a high flow of oxygen to filter and form urine. The vascular supply comprises a complicated network for distribution. The kidneys need a high flow of oxygen to filter and form urine. Glomerular capillary pressure is greater than in other organs and is optimal for filtration. Renal circulation has strong autoregulation. These mechanisms include myogenic mechanisms and tubuloglomerular feedback.

12.2.1.1 The Relationship with the Heart

From the heart through the abdominal aorta, the renal arteries directly supply the oxygen to the kidneys. Inside the kidney, at the hilium, are several branches, called interlobar arteries, that continue to the cortex. The glomerulus is then supplied by afferent arterioles, the subbranches of the interlobar arteries. Glomerular capillaries filter to the Bowman capsule and also form efferent arterioles that become peritubular capillaries around the renal tubules. These efferent arterioles are in the cortex within the medulla. The capillaries form venules; the veins then exit the kidneys as the renal vein.

- Myogenic mechanisms: a reduction in afferent arteriole pressure in smooth muscle for relaxation.
- Tubuloglomerular feedback: changes in pressure alter the filtration, flow and Na +, signaling dilation, or constriction. NO and prostacyclin may be involved in autoregulation. Renal vasoconstriction contributes significantly to systemic vascular resistance.

12.3 PART 3: KIDNEY DISEASES

Please go to the online course to find the answers.

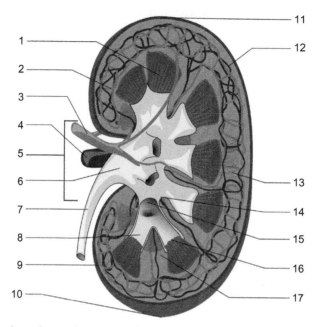

Image: from creative commons

Extrinsic factors	Intrinsic factors
Strong sympathetic stimulation response	-Very Weak metabolic control -Strong autoregulation

12.3.1 Balance and Interact

Emotional factors such as fear and cowardice cause the kidney qi to sink and fail to move upward, causing stagnation in the lower body.

12.3.2 Pathology
In pathology, the kidneys are affected by physiological factors. Kidney yin deficiency causes internal heat, leading to tinnitus, dizziness, hypersexuality, sore waist and weak knees, dry mouth, and nocturnal emission.

Diseases include:

- Hyperlipidemia
- Hypertension
- Diabetes insipidus

Prostatitis diseases include:

- Hyperlipidemia
- Hypertension
- Diabetes insipidus
- Prostatitis

Kidney yang uses heart fire to regulate kidney water, which regulates the heart fire. This relationship maintains the heart/kidney harmony. Kidney yang deficiency causes cold limbs, pain in waist and knees relieved by warmth, profuse clear urination, incontinence, and edema.

Diseases include:

- Diabetes 1 and 2
- Congestive heart failure
- Nephritis (chronic)
- Myocarditis (viral)

Kidney qi deficiency causes dyspnea with or without exertion, tachypnea, and other respiratory difficulties.

Diseases include:

- Asthma
- Bronchitis
- COPD
- Cor pulmonale

NOTES

For chapter 12 tutorial log on at www.niambiwellness.com to access the companion course for module 3.

CHAPTER 13

Liver Diseases

CHAPTER OBJECTIVES

After studying this chapter, you should be able to:

- Describe the generating and circulating properties of the liver zang.
- Describe the balancing and interacting properties of the liver zang.
- Explain the relationship between the liver and heart according to Western medicine.
- Explain the TCM and Western medicine perspectives in pathology.

13.1 PART 1: TRADITIONAL CHINESE MEDICINE

The liver stores blood as yin fluid. The surplus regulates volume and supply as needed to maintain circulation. The liver smoothes the free flow of qi, which also has a yang function of moving blood.

The functions of the liver are to:

- Store blood as a yin fluid
- Promote circulation of blood

13.2 PART 2: BASIC MEDICAL SCIENCE

13.2.1 The Relationship with the Heart

The liver is an important reservoir of blood. The network of vessels concerning the liver alone includes:

- Hepatic portal vein
- Hepatic artery

These arteries form sinusoids that filter blood throughout the liver. Liver circulation has weak sympathetic control, but sympathetic activation constricts both arteries.

Integrative Anatomy and Pathophysiology in TCM Cardiology. DOI: http://dx.doi.org/10.1016/B978-0-12-800123-3.00013-X

13.3 PART 3: LIVER DISEASES

Generate and circulate	Balance and interact
Strong for generating and circulating blood	Weak for psychoemotional balancing

13.3.1 Balance and Interact
Emotional factors such as anger injure the liver, causing the qi to rise and the blood to heat and ascend as well.

13.3.2 Pathology
In pathology, the liver is affected by physiological factors. Liver yin deficiency causes five palm heat; insomnia and dreaminess; dry skin, eyes and throat; and dizziness.

Diseases include:

- Coronary heart disease
- Hyperlipidemia
- Hypertension
- Menopause

Hyperactive liver yang causes rapidly rising anger, palpitations, irritability, dizziness, and a red face and eyes.

Diseases include:

- Coronary heart disease
- Hypertension

NOTES

For chapter 13 tutorial log on at www.niambiwellness.com to access the companion course for module 3.

CHAPTER 14

Spleen Diseases

CHAPTER OBJECTIVES

After studying this chapter, you should be able to:

- Describe the generating and circulating properties of the spleen zang.
- Describe the balancing and interacting properties of the spleen zang.
- Explain the relationship between the spleen and heart according to Western medicine.
- Explain the TCM and Western medicine perspectives in pathology.

14.1 PART 1: TRADITIONAL CHINESE MEDICINE

The spleen qi has the function of transforming food into nutrition for the production of blood, as well as keeping the blood contained in the vessels for normal circulation. The function of the spleen qi is to transform clear fluids and send turbid fluids to the kidneys. In turn, the kidneys resend the clear through the spleen to be used by the lungs.

Generate and circulate	Balance and interact
Strong for generating and transferring essence and generating blood	Weak for psychoemotional balancing

The functions of the spleen qi are:

- Transforming and transporting nutrition to generate qi and blood
- Controlling blood
- Transforming and transporting fluids

Integrative Anatomy and Pathophysiology in TCM Cardiology. DOI: http://dx.doi.org/10.1016/B978-0-12-800123-3.00014-1

14.2 PART 2: BASIC MEDICAL SCIENCE

14.2.1 Splenic Circulation

- The spleen is an important reservoir of blood filtering and production. It is supplied by the splenic artery, which branches from the celiac artery. The artery then drains into the portal vein.
- The relationship with the heart.
- The spleen does not have a direct participation in regular circulation but has a complicated open/closed intramicrocirculation. The spleen produces red blood cells, lymphocytes, and antibodies. Upon maturity, they are released to build the blood circulation and to help fight infections.

14.3 PART 3: SPLENIC DISEASES

Extrinsic Factors	Intrinsic Factors
Includes a strong sympathetic control when included in the gastrointestinal system	-Moderate metabolic control -No autoregulation

14.3.1 Balance and Interact

Over-thinking and lack of focus consumes the qi function of the spleen.

14.3.2 Pathology

In pathology, the spleen is affected by physiological factors. If blood flow to the brain is slow to block, the result is listlessness and unconsciousness. If spleen qi is deficient, then it fails in its transforming and transporting functions. One problem is that blood will be deficient.

Diseases include:

- Anemia
- Immune system problems

In addition, assisting the kidney yang in water balance and transport will be impaired.

Diseases include:

- Congestive heart failure
- Myocarditis (viral)

NOTES

For chapter 14 tutorial log on at www.niambiwellness.com to access the companion course for module 3.

CHAPTER 15

Gastrointestinal Diseases

CHAPTER OBJECTIVES

After studying this chapter, you should be able to:

- Describe the generating and circulating properties of the gastro fu.
- Describe the balancing and interacting properties of the gastro fu.
- Explain the relationship between the gastro fu organs and the heart according to Western medicine.
- Explain the TCM and Western medicine perspectives in pathology.

15.1 PART 1: CHINESE MEDICINE

Food is received by the stomach and processed into chyme to be later absorbed by the body. The small intestine receives the chyme and, working with the spleen, separates it into essence and waste product. The spleen qi helps to promote the digestion of food and the absorption of essence. The essence produced is sent to the lungs to be circulated as qi for the body. The free flow of liver qi produces bile, which is used in digestion. The bile is used to digest food processed by the stomach and the small intestine. Liver qi also helps the spleen function of transforming and transporting. The large intestine helps the kidney and spleen by receiving the turbid water and continues with its process of fluid reabsorption. Later, stool is discharged.

Generate and Circulate	Balance and Interact
Strong for generating and transferring essence and circulating blood.	-Weak for psychoemotional balancing

Copyright © 2014 Anika Niambi Al-Shura. Published by Elsevier Inc. All rights reserved.

The relevant organs of the gastro system:

- Stomach
- Small intestine

Integrative Anatomy and Pathophysiology in TCM Cardiology. DOI: http://dx.doi.org/10.1016/B978-0-12-800123-3.00015-3

- Spleen
- Liver
- Large intestine

The functions of the stomach qi:

- Receiving and processing food
- Dredge and descend

The functions of the small intestine:

- Digesting processed food from the stomach
- Separating clear from the turbid

The functions of the spleen qi:

- Transforming and transporting food and nutrition
- Transforming and transporting fluids

The functions of the liver qi:

- Secreting bile used in digestion
- Promoting the transformation and transportation of stomach and spleen

The functions of the large intestines:

- Transforming waste into stool
- Discharging the stool

15.2 PART 2: BASIC MEDICAL SCIENCE

15.2.1 The Relationship with the Heart

The stomach does not directly participate in regular circulation. However, considerable site blood flow happens at meal times. The hormones and gastric acid work in various ways to process certain foods. Vasodilation during meal times is mediated by hyperosmality and NO.

The stomach circulation is supplied by:

- Aorta
- Gastric artery
- Left gastric artery
- Left gastro-omental artery
- Right gastro-omental artery

- Gastroduodenal artery
- Short gastric arteries

The small intestine is supplied by the superior mesenteric artery, which joins the splenic artery and then drains into the portal vein. The small intestine does not directly participate in regular circulation. However, considerable site blood flow happens at meal times.

The spleen is an important reservoir of blood filtering and production. It is supplied by the splenic artery, which branches from the celiac artery. The artery then drains into the portal vein.

The liver is an important reservoir of blood. The network of vessels concerning the liver alone includes:

- Hepatic portal vein
- Hepatic artery

These arteries form sinusoids which filter blood throughout the liver. Elevations in pressure can lead to edema and ascites.

The large intestines are supplied by the superior and inferior mesenteric arteries. The large intestine does not directly influence circulation but rather reflects the condition. The gastro circulation receives a large amount of cardiac output; systemic vascular resistance is caused by sympathetic stimulation.

Parasympathetic stimulation increases glandular secretions and metabolic mechanisms which involves blood flow, NO, and bradykinin.

Adverse conditions such as GERD, heartburn, and constipation are symptoms involving blood flow or part of differentials in diagnosing heart diseases.

Differentials and algorithms are used in Western medicine to rule out certain symptoms and presentations when diagnosing.

15.3 PART 3: GASTROINTESTINAL DISEASES

Extrinsic Factors	Intrinsic Factors
Includes a strong sympathetic control	-Includes metabolic control -No autoregulation

15.3.1 Balance and Interact

- Anxiety and worry damages stomach qi. Over-thinking and lack of focus consumes the qi function of the spleen.
- Sorrow, mania, grief, and pensiveness consume the qi of the small and large intestines.
- Emotional factors such as anger injure the liver, making the qi rise and the blood heat and ascend as well.
- Liver qi directly affects the function and state of the gastrointestinal system.

15.3.2 Pathology

In pathology, the gastro system is attached to the more emotional aspects concerning the heart. Qi deficiency within the gastro system is connected to liver qi stagnation. Liver qi stagnation is connected to adverse psycho-emotional well-being. The qi of stomach, spleen and the small and large intestines are collectively the gastro qi. The gastro qi has the primary action of transforming and transporting food, fluids and waste, as well as separating the clear from the turbid. Gastro qi deficiency causes gas bloating, cold heavy limbs, poor appetite, abdominal pain and distension, chest pain, acid regurgitation, vomiting, and short breath.

Differentials include:

- Heartburn
- Gastritis
- Dyspepsia
- Ulcers
- Prolapses
- Infections
- Infestations

NOTES

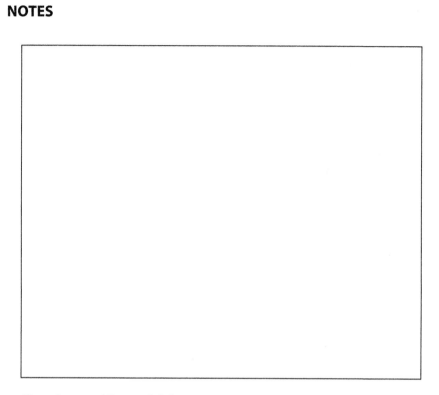

For chapter 15 tutorial log on at www.niambiwellness.com to access the companion course for module 3.

Module Review Questions

1. Describe the generating and circulating properties of the zang and fu organs.
2. Describe the balancing and interacting properties of the zang and fu.
3. Explain the interrelationships among the internal viscera, the brain, and the heart, according to Western medicine.
4. Explain the TCM and Western medicine perspectives in pathology.

Log on at www.niambiwellness.com to access the companion course and quiz for Module 3.

This also concludes the Integrative Anatomy and Pathophysiology in TCM Cardiology course. It is strongly suggested that you log onto the courses at the companion websites to review the course modules. Next, submit course documents and complete the final exam.

Upon passing the exam, you will receive completion certificates that include your name and practice license number, along with the specific number of credit hours awarded for this course. Electronic transmission of CEU and PDA credits will be sent to NCCAOM and your state medical board.

Printed and bound by CPI Group (UK) Ltd, Croydon, CR0 4YY

03/10/2024

01040421-0007